Consuming Cravings

A Transformative Approach to Recycling and Circular Economies for Zero-Waste Communities

Reba Redfern

Contents

1 Introduction to Zero Waste

Zero waste isn't about perfection; it's about progress. It's a journey, not a destination, a philosophy that encourages us to rethink our relationship with resources and minimize our environmental impact. This isn't about deprivation; it's about conscious consumption, making informed choices to reduce waste and live more sustainably. The core principle rests on refusing what we don't need, reducing what we do, reusing what we have, and recycling only as a last resort. It's about valuing resources and understanding their lifecycle, from creation to disposal, promoting circularity and minimizing waste at every stage. The benefits extend far beyond environmental protection. A zero-waste lifestyle contributes to economic growth by stimulating local businesses that support reuse and repair, fostering a more resilient and circular economy. It can also improve public health by reducing exposure to harmful toxins associated with landfills and incineration. Furthermore, adopting a zero-waste approach fosters a greater sense of community through shared experiences, collaborations, and initiatives like community composting programs. By actively participating in a more sustainable lifestyle, we improve not just the environment but our well-being, too.

Transitioning to zero waste involves a gradual shift in mindset and habits. Begin by assessing your current waste generation. What items end up in the trash the most? Identifying these patterns helps pinpoint areas where changes are most impactful. For example, food waste is often a significant contributor; strategies for better meal planning and food preservation can drastically reduce this type of waste. Similarly, single-use plastics are frequently problematic; choosing reusable alternatives like shopping bags, water bottles, and coffee cups makes a tangible difference. Start small, focusing on one area at a time, and gradually incorporate more zero-waste practices into your routine.

One misconception surrounding zero waste is that it demands extensive changes or specialized knowledge. In reality, even small adjustments make a significant difference. Switching from disposable to reusable items, such as napkins, straws, and containers, is a simple step with significant long-term impact. Composting food scraps and yard waste reduces landfill burden and creates nutrient-rich soil for your garden. Supporting local farmers' markets and businesses that prioritize sustainability promotes ethical consumption. These actions, seemingly small, contribute to a larger shift in our collective approach to waste.

The transition might feel daunting initially. It's a process that requires patience and self-compassion. Embrace imperfections; every effort counts. The journey itself offers valuable insights into consumption patterns, helping you develop mindful habits. It encourages a more meaningful connection with resources, fostering a sense of responsibility towards the environment. This transition isn't just about recycling; it's about challenging ingrained consumerist behavior and constructing new, environmentally conscious routines. It's about questioning the necessity of every purchase, and choosing to engage only in consumption that promotes environmental and social wellbeing. Remember that the essence of zero waste is not about achieving absolute zero waste – that's unrealistic and can feel unattainable. It's a mindset and a process that demands constant learning and improvement. It's about striving for continuous minimization of waste and maximizing the value and lifespan of resources. It's a collective endeavor, requiring collaboration, innovation, and policy changes to effectively address waste reduction. Embrace the journey, focus on consistent improvement, and appreciate every step forward. It's a journey that fosters personal growth, environmental responsibility, and a more sustainable future for all.

1.1 The Zero-Waste Philosophy

Zero-waste living isn't about perfection; it's a journey, a continuous striving towards minimizing our environmental footprint. It's about acknowledging the profound interconnectedness of our actions with the planet's health, recognizing that the seemingly insignificant act of discarding a plastic wrapper contributes to a larger, devastating problem of overflowing landfills and polluted ecosystems. This philosophy challenges the ingrained consumerist culture that prioritizes disposability and endless consumption, urging us to reconsider our relationship with resources. It prompts a deep dive into our habits, encouraging critical analysis and a determined shift towards conscious living. The core principle rests upon a conscious effort to refuse unnecessary items, embracing a minimalist lifestyle that values quality over quantity. Reducing consumption is paramount, focusing on needs over wants, and thoughtfully selecting products with longevity and repairability in mind. Reusing items creatively extends their lifespan, turning discarded materials into something new and useful, reducing the demand for virgin resources. Recycling, while imperfect, plays a supporting role—a last resort when refusal, reduction, and reuse are no longer viable options. Ideally, the waste that does eventually need processing is minimal, a testament to the effectiveness of prior stages in the zero-waste hierarchy.

Composting significantly reduces the organic waste stream, transforming food scraps and yard waste into nutrient-rich soil amendments. Home composting is surprisingly manageable, with various methods available depending on space and resources. A simple bin in the backyard or even a countertop composter can successfully handle household food waste. This process not only reduces landfill burden but also enriches the soil, promoting healthier plant growth and creating a closed-loop system where organic matter is perpetually recycled. Anaerobic digestion, a more sophisticated method, further reduces organic waste, converting it into biogas—a renewable energy source—and digestate, a valuable fertilizer.

Beyond individual actions, creating a zero-waste community demands collective effort. Advocating for policies that support waste reduction, promoting community composting initiatives, and participating in local clean-up efforts are essential steps. Educating others about the benefits of zero-waste living and the limitations of current recycling systems can significantly enhance broader adoption. Support local businesses committed to sustainable practices; choose products with minimal packaging; and prioritize durable, repairable goods. Remember, the collective impact of individual choices holds significant weight. The transition to zero-waste living requires patience and persistence. There will be setbacks and moments of frustration; it's a process of continuous learning and adjustment. Start with small, achievable goals; gradually incorporate new habits into your routine; don't be afraid to experiment; and most importantly, celebrate your successes along the way. Acknowledge that every small step counts, every conscious choice contributes towards a healthier planet. Focus on progress, not perfection. It is a lifestyle change, not a short-term project, that requires ongoing commitment and understanding.

Zero waste is more than just reducing trash; it's a fundamental shift in mindset, a conscious decision to live in harmony with our environment. It's about embracing a sustainable lifestyle that respects the planet's resources and acknowledges the interconnectedness of all living things. It's about consciously choosing longevity and durability over fleeting trends and disposable convenience. By adopting a zero-waste philosophy, we contribute to a more sustainable future, not only for ourselves but for generations to come. The journey may be challenging, filled with unexpected obstacles and moments of self-doubt, but the rewards—a cleaner environment, a healthier planet, and a more conscious way of life—are immeasurable. Remember, the collective power of individuals embracing this philosophy can fundamentally change the future, for a better tomorrow.

1.2 Benefits of Zero Waste Living

Embracing a zero-waste lifestyle offers a multitude of benefits extending far beyond simply reducing landfill waste. Firstly, and perhaps most significantly, it directly contributes to environmental protection. By minimizing our consumption and maximizing reuse, we lessen the strain on natural resources, reducing deforestation, pollution, and habitat destruction. This is not merely a feel-good notion; countless studies demonstrate the link between reduced waste and improved environmental quality. The reduction in greenhouse gas emissions from manufacturing, transportation, and decomposition of waste is another substantial benefit, a tangible contribution to mitigating climate change.

Furthermore, a shift towards zero-waste living fosters resource conservation. Instead of discarding items, we learn to repurpose and repair them, extending their lifecycles and conserving valuable raw materials. This inherently reduces the demand for new products, lowering the overall energy consumption associated with production and transportation. Consider the energy saved by mending a shirt rather than buying a new one; this seemingly small act, multiplied across countless individuals and items, becomes profoundly impactful. We become active participants in conserving our planet's resources, ensuring a more sustainable future for generations to come.

Beyond the environmental advantages, zero-waste living also brings considerable economic benefits. By reducing consumption and emphasizing reuse, we save money. Repairing items instead of replacing them, buying secondhand, and utilizing reusable containers all translate into significant long-term cost savings. Moreover, adopting a zero-waste lifestyle often leads to a more mindful approach to purchasing, preventing impulse buys and wasteful spending. The shift in mindset toward conscious consumption not only benefits the environment but also improves personal finances.

Beyond the tangible benefits, adopting zero-waste principles cultivates a deeper connection with our environment and communities. The process of reducing waste requires us to become more aware of our consumption habits and their impact. We learn to appreciate the resources we use and find creative ways to minimize waste. This heightened awareness can spark a sense of responsibility and empowerment, fostering a stronger connection to the planet. Moreover, community initiatives, such as shared composting systems and repair cafes, build stronger social bonds, fostering a sense of collective action toward a shared environmental goal. It's a profoundly rewarding experience, both personally and collectively.

Finally, adopting a zero-waste lifestyle is a powerful act of personal empowerment. By consciously choosing to reduce our environmental footprint, we take control of our impact on the planet, taking an active role instead of passively accepting the status quo. This sense of agency is empowering and instills a sense of purpose, fostering positive change from the ground up. The cumulative effect of many individuals making these conscious choices is substantial, creating a ripple effect of positive impact on the environment and society as a whole. It's not about perfection; it's about progress, and every step, however small, contributes to a healthier planet and a brighter future. This journey of mindful consumption is inherently rewarding, offering a sense of accomplishment and contribution that goes beyond mere convenience. Zero-waste living is a statement, a declaration of personal responsibility and environmental stewardship. It's a path toward a more sustainable and fulfilling life.

2 Understanding Consumption Habits

Our relationship with consumption is complex. It's interwoven with our identities, our social connections, and our very sense of well-being. We buy things to celebrate, to comfort ourselves, to impress others, and often, without even thinking. This seemingly simple act—the purchase of a product—has profound environmental consequences, impacting everything from resource depletion to pollution. Understanding this intricate dance between our desires and the planet's resources is crucial for transitioning towards a zero-waste future.

Let's start with a simple exercise: take a look at your trash. Really look at it. What do you see? Disposable coffee cups? Excessive packaging? Single-use plastics? The composition of your waste offers a powerful insight into your consumption habits. By meticulously tracking what you throw away for a week, even a single day, a clear pattern emerges revealing areas needing immediate change. You might be surprised by the sheer volume of items quickly discarded. This awareness, this honest self-assessment, forms the bedrock for conscious change.

Consumerism, fueled by relentless marketing and societal pressures, often encourages a cycle of acquiring, consuming, and discarding.

We're bombarded with messages suggesting happiness resides in the newest gadgets, the latest fashion trends, and the ever-expanding array of products vying for our attention. This cycle of planned obsolescence, intentionally designed for short lifespans, generates an enormous amount of waste, straining landfills and polluting ecosystems. The environmental cost is immense—resource depletion, habitat destruction, and the release of harmful toxins. The constant pressure to 'keep up' is exhausting and ultimately unsustainable.

Mindful consumption is not about deprivation; rather, it's a shift in perspective. It's about making conscious decisions based on need, value, and long-term impact. Before making a purchase, ask yourself: Do I really need this? Will this purchase contribute to my well-being or is it driven by fleeting desire? Are there sustainable alternatives? Could I borrow, rent, or repair instead of buying new? These simple questions can significantly alter shopping habits. Shifting from impulsive buying to thoughtful purchases takes practice.

Consider the longevity of a product. Will it last? Is it repairable? Choosing durable, high-quality items, even if they cost slightly more upfront, often proves more economical in the long run. By prioritizing quality over quantity, you reduce your overall consumption, minimize waste, and contribute to a more sustainable system. Investing in repairable items extends their lifespan, delaying the need for replacement and lessening the demand for new products.

This mindful approach extends beyond individual purchases to encompass a broader lifestyle shift. It involves seeking out businesses committed to ethical and sustainable practices, supporting local farmers markets, and choosing products with minimal packaging. It's about re-evaluating our priorities and understanding that true wealth lies not in accumulating possessions but in experiences, relationships, and environmental stewardship. The shift from a culture of consumerism to conscious consumption represents a fundamental change in societal values, prioritizing well-being over endless material acquisition. This transition may take time, but each small step counts.

The path toward a zero-waste lifestyle isn't a race but a journey of continuous learning and adaptation. Every conscious choice, no matter how small, contributes to the larger goal of reducing waste and fostering a more sustainable future. By understanding our consumption habits, we can begin to reshape our relationship with the environment, creating a more harmonious and sustainable world for ourselves and generations to come. This is not a solitary endeavor; it's a collective effort that requires individual commitment, systemic change, and ultimately, a global shift in consciousness. The journey towards zero waste is an investment in a healthier planet and a brighter future. The rewards—a cleaner environment, a more fulfilling lifestyle, and a stronger sense of community—are invaluable.

2.1 Analyzing Your Waste

Begin by honestly assessing your trash. Take a week, meticulously sorting your waste into categories: paper, plastic, glass, metal, food scraps, compostable materials, and finally, that frustrating "other" category – the things that don't easily fit anywhere else. Weigh each category if possible, or at least estimate the volume. This simple act of observation is surprisingly revealing. You might be surprised by the sheer volume of a specific waste stream, highlighting areas needing immediate attention. Don't just glance; truly examine the contents of each category; you'll find patterns emerge. This granular level of detail offers critical insight into your consumption habits.

Now, analyze those findings. What percentage of your waste is recyclable? How much is compostable? How much ends up in the landfill? The proportions will likely illuminate surprising truths about your daily routines. Perhaps you're consuming more single-use plastics than you realized, or your food waste constitutes a significant portion of your overall trash. Identify the largest categories. These are the targets for immediate action. Understanding the 'why' behind each item is crucial; is it convenience, lack of awareness, or something else entirely? This self-reflection forms the bedrock of meaningful change.

Focusing on your biggest waste culprits is key to immediate, visible progress. If single-use plastics dominate, start by consciously choosing reusable alternatives. Invest in a good quality reusable water bottle, coffee cup, and shopping bags. Pack your lunch in reusable containers instead of relying on disposable ones. Small changes accumulate; these are not monumental shifts, but a gradual transition to a more sustainable lifestyle. Small choices, made repeatedly, become powerful habits.

Food waste is a significant environmental problem. Plan your meals meticulously; make shopping lists and stick to them. Learn to properly store your food to extend its shelf life, and get creative with leftovers. Composting your food scraps is another powerful step you can take. It doesn't require complex systems; even a simple bin under your sink can make a difference. This diverts organic matter from landfills, reducing methane emissions and enriching your garden soil simultaneously.

Beyond the immediate changes, consider broader lifestyle shifts. Examine your purchasing habits. Do you buy things impulsively? Do you prioritize quality over quantity? Prioritizing durable, repairable items extends their lifespan, significantly reducing your overall waste. Support businesses that align with your values; those that prioritize sustainability and minimize packaging, for instance. This extends beyond individual action; it impacts the production and consumption chain at a much larger level.

Don't be discouraged by setbacks. Transitioning to a zero-waste lifestyle is a journey, not a race. You will inevitably slip up; everyone does. The key is to learn from your mistakes and keep moving forward. Remember the initial analysis; revisit it regularly. Track your progress over time; use a journal or a spreadsheet to chart your waste reduction efforts. Seeing tangible progress can boost your motivation and reaffirm your commitment.

Remember, the goal is not perfection; it's progress. Every small step, every conscious choice, contributes to a larger movement towards a more sustainable future. Your individual journey is a significant piece of this larger puzzle. Analyzing your waste is not just an exercise in data collection, it's an introspective journey that reveals patterns in your consumption habits, allowing you to make conscious and sustainable changes that benefit both you and the planet. The power lies in understanding your waste, acknowledging your impact and choosing to consume more mindfully. It's a process of constant learning, refinement, and commitment to a more sustainable future.

2.2 Consumerism's Impact

Consumerism, fueled by relentless advertising and readily available credit, fosters a culture of disposability. We buy, we use, we discard, often without a second thought to the environmental consequences. This cycle, repeated billions of times daily, generates mountains of waste, depletes natural resources, and contributes significantly to climate change. The sheer volume of packaging alone is staggering, a testament to a system prioritizing profit over planet. The impact extends beyond landfills; the extraction of raw materials, manufacturing processes, and transportation all leave their carbon footprint. Our insatiable desire for the newest gadgets, the latest fashion trends, and fleeting conveniences, is fundamentally unsustainable. This isn't about shaming individual choices; it's about acknowledging the systemic issues that make conscious consumption so challenging.

The economic model itself is geared towards continuous growth, demanding ever-increasing consumption to maintain profitability. This relentless pursuit of economic expansion ignores the finite nature of our planet's resources. Planned obsolescence, where products are designed to break down or become outdated quickly, further exacerbates the problem. This isn't simply a matter of faulty design; it's a deliberate strategy employed to stimulate demand and boost sales. Consider the countless electronic devices discarded each year, often containing valuable and recoverable materials. The environmental and social costs of mining these resources, often in developing countries, are rarely factored into the final price tag. This disregard for true cost creates an illusion of affordability, masking the hidden environmental and social debt we accumulate.

Understanding the broader implications is crucial. The production of goods consumes vast amounts of energy, contributing to greenhouse gas emissions and global warming. The disposal of these goods, often in uncontrolled landfills, leads to soil and water contamination, harming ecosystems and human health. The extraction of raw materials often involves habitat destruction, biodiversity loss, and displacement of communities. These are not isolated incidents; they are interconnected aspects of a system deeply rooted in unsustainable practices. We need to move beyond individual responsibility and address the structural problems driving this cycle of consumption and waste.

This requires a fundamental shift in our thinking, moving away from a linear "take-make-dispose" model to a circular economy. A circular economy aims to minimize waste and maximize resource utilization by designing products for durability, repairability, and recyclability. It emphasizes reuse, sharing, and repair over constant consumption. This transition will require changes at every level—from manufacturers designing more sustainable products to consumers making more conscious purchasing decisions. Governments have a crucial role to play in creating policies that incentivize sustainable practices and discourage unsustainable ones. These policies could range from extended producer responsibility schemes, holding manufacturers accountable for the end-of-life management of their products, to regulations on packaging and waste disposal.

However, systemic change is not enough. Individual actions remain vital. By consciously reducing our consumption, choosing durable and repairable products, and embracing reuse and repair, we can significantly lessen our environmental impact. We must cultivate a deeper appreciation for the value of goods, resisting the allure of fleeting trends and prioritizing quality over quantity. Support businesses committed to ethical and sustainable practices; actively seek out products made from recycled materials or designed for longevity. These small actions, when multiplied across a community, can create significant change. Understanding consumerism's impact isn't about guilt; it's about empowerment – empowering ourselves and our communities to build a more sustainable future. It is about recognizing that our purchasing power can be a force for positive change, driving the shift towards a more responsible and sustainable economic model. The journey towards zero waste begins with a critical examination of our own consumption habits, recognizing the far-reaching implications of our choices and actively working towards a more sustainable lifestyle.

2.3 Mindful Consumption

We've explored the impact of consumerism and the importance of reducing waste. Now, let's delve into the heart of the matter: mindful consumption. This isn't about deprivation; it's about making conscious choices. It's about questioning every purchase, considering its lifecycle, and prioritizing quality over quantity. Think about the clothes you buy: Do you really need another T-shirt, or could you mend an old one? That seemingly small decision, repeated countless times, drastically alters your environmental footprint.

Mindful consumption starts with awareness. Before you buy something, ask yourself: Do I truly need this? Will this item serve a purpose beyond fleeting satisfaction? Can I borrow or rent it instead? Is there a secondhand alternative? These simple questions, when consistently applied, can reshape your shopping habits. Consider the packaging: Is it recyclable or compostable? Are there less wasteful alternatives available? By actively considering these questions, you take control of your consumption and actively contribute to a circular economy.

Next, consider the longevity of products. Choosing durable, high-quality items that are built to last dramatically reduces the frequency of replacements. This applies to everything from clothing and kitchenware to electronics and furniture. Look for products made from sustainable materials or with minimal packaging. Repairing items instead of discarding them is another vital component. Learning basic repair skills for clothing, electronics, or furniture can extend the life of your possessions, saving money and reducing waste. There's a wealth of online resources and community workshops available to help you acquire these skills. Remember, this isn't just about individual actions; supporting businesses that prioritize sustainability is equally crucial. Patronize companies that use sustainable materials, ethical labor practices, and minimal packaging.

Furthermore, the concept of mindful consumption extends beyond material goods. Consider your digital consumption – the apps you use, the streaming services you subscribe to, and the sheer volume of data you generate. These aspects, often overlooked, contribute significantly to our collective environmental impact. By being mindful of your digital footprint, you can reduce energy consumption and minimize e-waste. This extends to the services you use; do you really need all those streaming subscriptions? Could you borrow a book instead of buying it? Consider the hidden costs of consumption, such as transportation and manufacturing. Often, items seemingly cheap at the point of purchase carry a far greater environmental price tag than is immediately apparent.

Finally, mindful consumption is an ongoing journey, not a destination. There will be times when you slip up, when you make impulsive purchases. That's okay. The key is to learn from those mistakes, to reflect on your buying habits, and to continuously strive for greater awareness and intentionality. It's about progress, not perfection. By integrating mindful consumption into your daily life, you're not merely reducing your waste; you're actively participating in the creation of a more sustainable and equitable future. It's about actively engaging with your consumption habits, being aware of their impact, and making conscious choices to minimize your environmental footprint. This process should be iterative, allowing for continuous learning and adaptation. The journey towards mindful consumption is a personal one, and its rewards are both individual and collective. You'll save money, reduce clutter, and feel a sense of accomplishment knowing you're contributing to a healthier planet. This mindful approach is integral to achieving a zero-waste lifestyle, not as a strict regimen, but as a philosophy guiding your interactions with the world.

3 Recycling and its Limitations

Many believe recycling is the ultimate solution to our waste problem; a simple act of tossing a bottle into a blue bin magically transforms it into something new. This is far from the truth. The reality is significantly more nuanced, demanding a deeper understanding of the process and its inherent limitations. Recycling, while valuable, is not a silver bullet.

The first misconception to debunk is the idea that all materials are easily recyclable. Many plastics, for example, are incredibly difficult and costly to recycle, often ending up in landfills or incinerators. Even seemingly straightforward materials like paper and cardboard have limitations, requiring specific processing and often resulting in a lower-quality product after recycling. The energy and resources consumed during the recycling process itself must also be considered; sometimes, it's more efficient to create new products from raw materials.

Downcycling, a frequent consequence of recycling, further complicates matters. While recycling turns waste into something usable, downcycling creates a product of lower quality than the original. Think about a plastic bottle becoming a fleece jacket: the jacket isn't as robust or high-quality as the original bottle. Through multiple cycles of downcycling, the material eventually becomes unusable, inevitably ending up in a landfill. This continuous degradation underscores the need for more radical solutions beyond simple recycling.

Improving our recycling systems is paramount to maximizing their effectiveness. This necessitates increased investment in advanced sorting technologies, capable of separating materials with greater precision. Additionally, a shift towards standardized recycling protocols across regions would significantly enhance efficiency and reduce confusion. Cleaner waste streams, free from contamination, are crucial for efficient recycling. Public education campaigns that clearly explain these issues and provide practical guidance on sorting materials are essential for success.

Composting provides a crucial alternative for organic waste, transforming food scraps and yard waste into nutrient-rich soil amendments. This process not only diverts organic matter from landfills but also reduces greenhouse gas emissions and creates a valuable resource for gardeners and farmers. Home composting is accessible, manageable, and an effective method for reducing kitchen waste. However, for larger communities, industrial composting facilities, employing advanced technologies, may be necessary to handle the volume of organic waste effectively.

Anaerobic digestion, another efficient method, processes organic waste in the absence of oxygen, producing biogas—a renewable energy source—and digestate—a valuable fertilizer. Unlike composting, this process is best suited for large-scale operations, requiring specialized infrastructure and expertise. Anaerobic digestion offers a holistic solution by generating energy and fertilizer while diverting waste from landfills. While the initial investment can be high, the long-term benefits, both economically and environmentally, are significant. Considering the feasibility of establishing local anaerobic digestion plants should be a priority in community waste management plans.

Ultimately, a comprehensive approach to waste management requires a move beyond solely relying on recycling. Understanding its limitations and incorporating alternative methods like composting and anaerobic digestion are crucial for creating truly sustainable waste management systems. A multi-pronged approach, focusing on reducing waste generation at its source, combined with effective reuse and recycling strategies, offers the best pathway toward a zero-waste future. This requires individual commitment, innovative technologies, and supportive policies to succeed.

3.1 Recycling Myths Debunked

Many believe that simply placing items in a designated recycling bin guarantees their transformation into new products. This isn't always the case. Recycling infrastructure varies significantly across regions, and contamination – even a single non-recyclable item in a bin of recyclables – can render an entire load unusable. The reality is much more nuanced and often involves sorting, processing, and sometimes, disposal, highlighting the limitations of our current recycling systems.

Wishful thinking often paints a picture of a perfectly efficient system, where every plastic bottle is reborn as another. However, the process is far from perfect. Many plastics, for example, can only be downcycled—meaning they're transformed into lower-quality products after each recycling cycle, eventually becoming unrecyclable. Similarly, the energy required for recycling, including transportation and processing, can outweigh the environmental benefits, especially for items with low recycling rates. Understanding these realities fosters a more responsible approach to waste management.

We often hear that recycling "saves the planet," fueling the perception that recycling everything is equally effective. This isn't accurate. Recycling aluminum cans is far more environmentally friendly than recycling many types of plastics due to the comparatively high recycling rates and lower energy consumption associated with aluminum processing. Focusing our efforts on materials with higher recycling rates and lower environmental impact maximizes the positive effect of our recycling actions. Prioritizing reduction and reuse before recycling is a significantly more effective strategy for minimizing environmental impact.

The idea that all plastics are recyclable is pervasive, yet misleading. The diverse range of plastics, each with unique chemical compositions and properties, makes comprehensive recycling challenging. Many plastics simply aren't economically feasible to recycle due to complex separation processes, costly material recovery, or lack of demand for recycled materials. Focusing on reducing our consumption of single-use plastics and opting for reusable alternatives is a more sustainable solution than solely relying on the current recycling infrastructure.

The common misconception that "a little contamination is okay" is a dangerous one. Contamination can disrupt the entire recycling process. A single greasy pizza box or a plastic bag in a bin of paper can compromise an entire load, making it impossible to process effectively and resulting in the entire batch being sent to landfill. Meticulous sorting and cleaning are crucial to maximize recycling efficiency, and it is vital to understand which materials are accepted for recycling in your specific area to avoid contamination.

Furthermore, some materials, even when cleanly sorted, aren't always recyclable in your local system. Local recycling guidelines, frequently overlooked or misunderstood, greatly influence the viability of recycling efforts. Your local municipality's recycling guidelines are critical; adhering to them ensures proper processing and prevents contamination. A quick visit to your local government website or a call to your waste management provider will clarify which materials are acceptable. Ignoring this leads to misplaced recycling efforts and increased landfill waste.

The myth of infinite recyclability is a particularly harmful one. Most materials lose quality with each recycling cycle, eventually becoming unsuitable for recycling. This is especially true for many types of plastics which are often downcycled. This reality challenges the assumption that endlessly recycling a material is sustainable in the long term. Strategies to reduce the initial production and use are far more critical than relying on continuous cycles of imperfect recycling processes. A mindset focused on reducing consumption before recycling is essential.

The narrative that recycling solves the waste problem entirely is fundamentally flawed. Recycling is just one element of a multifaceted solution. It's crucial to incorporate other waste management strategies, such as reduction, reuse, composting, and anaerobic digestion, to achieve significant progress towards zero-waste goals. A holistic approach that prioritizes waste prevention is considerably more effective than solely relying on recycling to address the issues of waste generation and environmental impact. Focusing our energies on waste reduction and reuse, while understanding the limitations of recycling, ensures a more efficient and effective approach.

3.2 The Downcycling Problem

Downcycling, a process where recycled materials are converted into products of lesser value, presents a significant challenge to achieving true circularity. Unlike recycling, which ideally aims to create products of equal or greater value, downcycling leads to a gradual depletion of resources. Think about plastic bottles: recycled into fleece, its value is undeniably diminished. This continuous degradation hinders the creation of a truly closed-loop system, perpetuating the reliance on virgin materials and intensifying environmental problems. The inherent limitations of downcycling become strikingly apparent when considering the energy and resources consumed in the process, often outweighing the benefits achieved.

The problem is further compounded by the complexities of material sorting and contamination. Many plastics, for instance, are difficult to identify and separate effectively, leading to mixing of incompatible materials and rendering them unsuitable for proper recycling, thereby pushing them towards downcycling. This issue is not restricted to plastics; various materials face similar challenges. Furthermore, the market demand for downcycled products often lags, resulting in lower economic incentives for the recycling industry and contributing to the prevalence of landfilling. Consequently, a considerable amount of potentially recyclable material ends up in landfills, thereby contributing to environmental pollution and harming ecosystems.

We need to move away from a reliance on downcycling and strive towards true closed-loop systems. One crucial step is designing products with recyclability in mind. This involves using materials that are easily separable and recyclable, as well as avoiding the use of mixed materials which complicate the recycling process. Consider using single-material packaging, a design choice that vastly simplifies the sorting and processing stages. This proactive approach is vital in ensuring materials can be easily recycled into products of equivalent or even higher value.

Investing in advanced recycling technologies offers another path towards a more circular economy. Chemical recycling, for example, can break down plastics into their basic building blocks, allowing for the creation of new, high-quality materials. These technologies are still in their development phases, however, and their widespread adoption requires substantial investment in research and infrastructure. Furthermore, significant challenges remain in addressing the scale of production and the economic viability of these methods. But if scaled effectively, these innovative approaches hold the key to circumventing the downcycling trap and moving closer towards true material circularity.

The issue isn't simply technological; consumer awareness and behavior play a crucial role. Educating consumers about the limitations of downcycling and the importance of choosing products designed for recyclability is critical. This shift in consumer demand will drive innovation and investment in higher-quality recycling infrastructure and technologies. Moreover, implementing clear and consistent labeling systems for recycled content can provide consumers with the necessary information to make informed purchasing decisions. This informed consumer base will eventually push manufacturers toward adopting

more sustainable practices and incorporating recyclable, reusable materials into their products.
Ultimately, overcoming the downcycling problem requires a multipronged approach. It calls for collaborative efforts between governments, industries, and consumers. Stringent regulations and policies can help incentivize the development and implementation of advanced recycling technologies. Public awareness campaigns are needed to dispel common myths surrounding recycling and to empower consumers to make conscious choices that support circular economy initiatives. By embracing innovation, fostering collaboration, and promoting responsible consumption, we can transition towards a more sustainable future where the downcycling problem is effectively mitigated, and true circularity becomes a reality. The transition necessitates a comprehensive shift in how we design, produce, consume, and manage materials, leading us towards a more sustainable and efficient system. Only through these combined efforts can we fully realize the potential of a circular economy and minimize the negative impacts of downcycling on our planet.

3.3 Improving Recycling Systems

Our current recycling systems, while well-intentioned, often fall short of their potential. Many municipalities struggle with inconsistent collection methods, leading to contamination and ultimately, landfill-bound materials. The lack of standardized sorting protocols across different regions creates further challenges. A significant portion of supposedly recyclable materials ends up incinerated or in landfills due to these inconsistencies and limitations. Addressing these issues requires a multi-pronged approach.

Improving recycling infrastructure is paramount. This involves investing in advanced sorting technologies, such as automated sorting systems equipped with near-infrared (NIR) spectroscopy and artificial intelligence for material identification. These systems can accurately sort materials with much higher precision than manual sorting, minimizing contamination and maximizing recovery rates. Furthermore, improved public education campaigns are crucial. Clear, concise, and easily accessible information about what materials are accepted for recycling within a specific region is essential for increasing participation rates and reducing contamination. The information must be readily available via diverse channels – websites, brochures, mobile apps, and community outreach programs.

We need to rethink the design of recyclable products. Manufacturers should prioritize using easily recyclable materials and designing products for disassembly and material recovery. The shift towards designing for recyclability requires collaborative efforts between designers, manufacturers, and recyclers. This means adopting standardized materials and designs, ensuring compatibility with existing recycling systems, and minimizing the use of mixed materials, which are difficult and expensive to separate. Such design innovations would drastically enhance recycling effectiveness and streamline the entire process.

Beyond material improvements, we must consider the economics of recycling. Currently, the low value of recycled materials often makes recycling less economically viable than landfilling or incineration. Government subsidies and extended producer responsibility (EPR) schemes could help bridge this economic gap. EPR programs hold manufacturers accountable for the end-of-life management of their products, incentivizing them to design products for recyclability and to contribute financially to recycling infrastructure. This shift in responsibility, from the consumer to the producer, would make recycling a more financially attractive proposition.

Expanding the range of recyclable materials is also crucial. Many materials that are currently considered unrecyclable, such as certain plastics and textiles, possess potential for recovery and reuse. Research and development into new recycling technologies, like chemical recycling processes, could unlock this potential and offer novel solutions. This would require investment in research, development, and the deployment of these innovative technologies at scale, creating new opportunities for green jobs and economic growth. Public-private partnerships, fostering collaborative efforts among governments, industries, and research institutions, are essential to accelerate the process.

Finally, we cannot ignore the role of consumer behavior. Increased consumer awareness and responsible waste disposal habits are crucial for the success of any recycling program. Education campaigns emphasizing the importance of sorting materials correctly, avoiding contamination, and reducing overall waste are essential. The focus should be on empowering consumers to become active participants in the recycling process rather than simply passive contributors. Gamification techniques and rewards programs could encourage greater participation and foster a sense of community engagement around waste reduction and recycling. The goal is to cultivate a culture of responsibility and environmental stewardship, where recycling becomes a collective endeavor. Ultimately, a concerted effort encompassing technological advancements, policy changes, and public awareness is required to optimize our recycling systems and create a truly sustainable future.

3.4 Composting Basics

Composting transforms organic waste—food scraps, yard trimmings—into a valuable soil amendment, diverting it from landfills and reducing methane emissions, a potent greenhouse gas. It's a simple process, readily adaptable to various lifestyles and spaces, from small apartment balconies to sprawling backyards. Think of it as nature's recycling system, accelerating decomposition and enriching your garden.

Successfully composting relies heavily on maintaining the right balance of "browns" and "greens." Browns are carbon-rich materials like dried leaves, twigs, shredded paper, and cardboard. Greens are nitrogen-rich materials, including grass clippings, fruit and vegetable scraps, coffee grounds, and tea bags. Aim for a roughly 2:1 ratio of browns to greens; too many greens can lead to unpleasant odors and attract pests. Too many browns result in a slow decomposition process.

The composting process itself happens through microbial activity. Microbes, tiny organisms, break down the organic matter, releasing heat in the process. This creates an ideal environment for the decomposition to take place efficiently. You'll notice the pile heating up as the microbes work. Turning the compost pile regularly, about once a week, aerates it, supplying oxygen to the microbes and speeding up the decomposition. This turning also helps to distribute moisture evenly throughout the pile. A consistently moist pile, like a wrung-out sponge, is crucial; too dry and decomposition slows, too wet and it becomes anaerobic, creating unpleasant odors.

Several composting methods exist. For those with limited space, a worm composting bin, or vermicomposting, offers a compact and efficient solution. Worms consume the organic matter, producing nutrient-rich castings, which are essentially worm poop, an excellent fertilizer. Alternatively, a compost tumbler provides easy access for turning and protects the pile from the elements. Larger setups, like backyard compost piles, allow for greater volume but require more space and often involve more manual labor. Regardless of the method, ensuring proper aeration and moisture levels remain key to success.

Regardless of your chosen method, you should monitor your compost regularly. Look for signs of healthy decomposition: dark brown color, earthy smell, and a crumbly texture. If the pile is too wet, add more browns. If it's too dry, add water. If unpleasant odors develop, it's likely too wet or lacks sufficient browns. Addressing these issues promptly prevents the pile from becoming a breeding ground for unwanted pests or developing foul-smelling anaerobic conditions.

Remember that composting isn't a magic bullet. Some materials should be avoided. Dairy products, meat, oily foods, and pet waste attract pests and can create foul-smelling conditions. These items attract unwanted animals and cause undesirable odors. Diseases can also spread easily through them, damaging the compost pile itself. Always err on the side of caution; it's better to leave these items out of your compost pile.

Once your compost is fully decomposed, it's ready to enrich your garden. This nutrient-rich material provides essential nutrients to plants, improving soil structure and promoting healthy growth. Use it as a soil amendment, incorporating it into your garden beds or potting mix. You'll find that your plants thrive thanks to the abundance of nutrients in your homemade compost. Compost helps retain water in the soil, thus reducing the need for frequent watering, saving you both

time and water resources.
Beyond personal use, composting offers significant environmental benefits. By diverting organic waste from landfills, it reduces greenhouse gas emissions, conserves landfill space, and contributes to a more sustainable waste management system. It's a simple act with far-reaching consequences, promoting a circular economy and reducing our overall environmental footprint. Consider joining a community composting program or starting a shared compost with neighbors to increase efficiency and participation. By actively participating in composting, you directly contribute to a healthier planet and a more sustainable future. This contribution, in turn, strengthens your community and enhances environmental stewardship overall. This small action ripples outward, creating positive change across multiple scales.

3.5 Anaerobic Digestion

Anaerobic digestion offers a powerful solution for managing organic waste streams, transforming them into valuable resources. Unlike composting, which relies on aerobic (oxygen-rich) conditions, this process uses microorganisms in an oxygen-free environment to break down biodegradable materials. This breakdown produces biogas, a mixture primarily of methane and carbon dioxide, which can be harnessed for energy generation, heating, or even transportation fuel. The remaining digested material, known as digestate, is a nutrient-rich soil amendment, valuable for agriculture and horticulture, effectively closing the loop on waste management.
The process itself is relatively straightforward. Organic waste, which

may include food scraps, agricultural residues, sewage sludge, and other similar materials, is first collected and pre-treated. This pretreatment step may involve size reduction, screening, or homogenization, to ensure efficient digestion. The pre-treated material is then placed in a sealed anaerobic digester, where microorganisms, predominantly anaerobic bacteria, begin the decomposition process. Temperature and pH are carefully monitored and controlled to optimize digestion efficiency and biogas production. These controlled conditions are key; anaerobic digestion requires a careful balance of many factors for the process to work correctly.

Different digester designs exist, each suited to varying scales and waste streams. Large-scale industrial digesters are typically employed for processing substantial quantities of waste. These systems may incorporate multiple stages or use advanced technologies to maximize biogas yield and optimize digestate quality. Smaller-scale systems, such as those used in rural communities or individual farms, provide a more localized approach to waste management, creating a more community-centred method of treating local waste. Choosing the appropriate digester size and design requires careful consideration of factors such as waste volume, available land, and desired biogas output. This selection requires a deep understanding of the local context and waste generation dynamics, and it can often make the difference between success and failure of the entire project.

Beyond the technological aspects, successful anaerobic digestion relies on effective waste collection and sorting systems. Contaminants like plastics, metals, or chemicals can inhibit microbial activity and reduce biogas production. Therefore, ensuring a clean, consistent feedstock is crucial for the long-term success of an anaerobic digestion facility. This requires community education and participation, building awareness about acceptable waste materials and the importance of proper separation. Effective community engagement is paramount, highlighting the role of education in ensuring consistent high-quality input materials. This isn't just about teaching people what to put in the bin, but about building a shared understanding of the environmental impact of their waste disposal choices.

The potential benefits of anaerobic digestion extend beyond waste management. By generating renewable energy, it reduces reliance on fossil fuels, contributing to a cleaner, more sustainable energy mix. The digestate, as a valuable fertilizer, can replace synthetic fertilizers, reducing environmental impacts associated with their production and transport. It's crucial to note that the digestate, while nutrient-rich, needs to be handled and applied correctly to avoid any potential environmental harm. It is of paramount importance to correctly implement the technology in the local context and to respect local regulations. Furthermore, the cost-effectiveness of anaerobic digestion depends largely on factors like waste availability, energy prices, and policy support. Government incentives, such as subsidies or tax breaks, can significantly enhance its economic viability, and are often necessary to make it a feasible solution.

In conclusion, anaerobic digestion presents a compelling solution for communities striving towards zero-waste goals. It offers a sustainable approach to managing organic waste, transforming it into renewable energy and valuable resources. However, the successful implementation requires a holistic approach, integrating technical expertise, effective waste management systems, and proactive community engagement. Thorough planning and consideration of local conditions are imperative for the efficient and effective operation of this process. Proper management and integration within the local regulatory framework is crucial for its sustainability and for its positive impact on the environment. Only through a careful consideration of these various aspects can this powerful technology truly deliver on its considerable promise.

4 Circular Economy Principles

Designing products for durability is paramount. We must move beyond the disposable mentality that fuels our current linear economy. Think about the clothes you wear, the appliances in your home, even the packaging of your groceries. How long do they truly last? How easily can they be repaired? A circular economy demands that we design products with longevity in mind, incorporating repairable components, modular designs, and readily available parts for easy fixes. This requires a fundamental shift in manufacturing processes and design thinking, prioritizing quality and longevity over planned obsolescence. This isn't just about extending the lifespan of individual items; it's about creating a system where resources are valued and reused, preventing the constant cycle of extraction, production, and disposal. The benefits extend far beyond the simple avoidance of waste, influencing everything from economic productivity to resource security and environmental protection. Companies should be incentivized to create items meant to last, leading to sustainable business practices. This shift should be reflected in design choices, using materials that are easily recycled or repurposed at the end of a product's life. Consumer awareness

plays a vital role too. By demanding durable and repairable products, consumers can signal the market demand for long-lasting goods.
Product lifecycle management (PLM) offers a systematic approach to minimize environmental impact throughout a product's entire journey. From conception to disposal, every stage demands careful consideration. Effective PLM begins with responsible sourcing of raw materials, favoring sustainable and ethically harvested resources. The manufacturing process itself must be optimized for efficiency and minimal waste generation. The design phase should prioritize recyclability and the use of renewable materials whenever possible. Careful attention needs to be paid during the product's use phase, focusing on reducing consumption and promoting repair and maintenance rather than instant replacement. Finally, end-of-life management, including recycling, reuse, and responsible disposal, is integral. Employing PLM systems allows businesses to track resources, identify waste streams, and continuously improve their environmental performance. Investing in robust data management systems helps firms analyze product performance data, optimize manufacturing processes, and understand the environmental footprint of their products across their whole lifecycle, which helps to improve both sustainability and profitability. This data-driven approach allows for continuous improvement, enhancing both the efficiency and environmental performance of the product lifecycle. The potential for improvement is substantial, with significant environmental and economic benefits resulting from a comprehensive PLM strategy.
The circular economy requires a fundamental shift in our thinking – from a linear "take-make-dispose" model to a cyclical system where resources are kept in use for as long as possible, extracting maximum

value before responsibly managing their end-of-life. This transition necessitates a multifaceted approach, involving not only innovative technologies and business models but also a significant change in societal attitudes and consumption patterns. Designing for durability and implementing effective product lifecycle management are pivotal aspects of the process. It's about a holistic approach that considers all elements of the system and promotes collaboration across sectors. By integrating these key principles, we can begin moving toward a more sustainable future, one where our consumption patterns don't deplete resources or harm the planet. It requires collective action and a shared responsibility across industries, communities, and consumers to embrace circular economy principles and contribute to a better future for generations to come. The adoption of circular economy models also brings significant economic opportunities. A shift from waste disposal to resource recovery creates new industries, employment opportunities, and innovative business models. This includes waste processing and recycling facilities, the development and implementation of repair and reuse services, and the creation of new products from recycled materials. The transition to a circular economy can lead to enhanced resource security, reducing dependence on finite resources and strengthening supply chains. This resilience is vital in an era of increasing resource scarcity and global uncertainty.

Collaboration is key. Businesses, governments, and communities must work together to create supportive policies, infrastructure, and market conditions that incentivize the adoption of circular economy practices. This requires investment in research and development to drive innovation, creating technological solutions to challenges like materials recycling, product design, and waste management. It also entails fostering a culture of innovation through collaboration between industry, academia, and research institutions, creating and sharing best practices, and developing standardized approaches to assess and measure progress towards circularity. Government support includes regulations and incentives to promote sustainable practices, including tax breaks for companies implementing circular economy models, and funding for research and development in areas like waste management, sustainable materials, and product design. Policies could include extended producer responsibility schemes, where producers are held responsible for the entire lifecycle of their products, incentivizing the design of products for durability, recyclability, and repairability. Raising public awareness through education and communication is just as crucial. This includes informing consumers about the benefits of circular economy practices, making them aware of how to participate, and changing their consumption habits. Education, particularly in schools, can promote greater awareness and encourage the next generation to embrace sustainable practices. Only through a collaborative, multifaceted approach can we successfully transition to a circular economy.

4.1 Designing for Durability

Durability isn't just about a product lasting longer; it's about designing for a future where resources are valued, not discarded. We need to move beyond the throwaway culture, demanding products that stand the test of time, both physically and functionally. This means considering the entire lifecycle of a product, from its conception to its eventual decomposition. Think about the materials chosen, their sourcing, and their potential for reuse or recycling. A truly durable product is designed with its end-of-life in mind, minimizing its environmental footprint.

Designing for longevity necessitates a shift in mindset. We must prioritize quality over quantity, opting for fewer, well-made items rather than a constant stream of cheap, disposable goods. This isn't about austerity; it's about investing in products that deliver lasting value. Consider the emotional connection we develop with durable goods; they become cherished possessions, extending their lifespan far beyond their initial purpose. This contrasts sharply with disposable items, often quickly forgotten.

Repair and maintenance should not be afterthoughts, but integral parts of a product's design. Products should be easy to repair, with readily available parts and clear instructions. Modular designs, where components can be easily replaced, are particularly valuable. This approach extends the product's lifespan, reduces waste, and empowers consumers to become active participants in the product's lifecycle. Imagine a world where broken appliances aren't simply thrown away but easily fixed, with readily available parts and clear, straightforward instructions.

Consider the materials used. Opting for sustainable and recyclable materials is crucial. Biodegradable materials, while offering a solution for the eventual decomposition of the product, require careful consideration of their own lifecycle and potential impact. Choosing materials that are both durable and easily recyclable ensures the product's components can be reintroduced into the circular economy, minimizing waste and resource depletion. The environmental impact of sourcing and processing materials must always be a primary concern in the design process.

Furthermore, the design process itself should be conscious of its impact. Consider employing sustainable manufacturing practices, minimizing energy consumption and waste generation during production. Ethical sourcing of materials is also paramount, ensuring that the production process is fair and equitable to workers and communities. The design phase is not just about aesthetics and function; it's an opportunity to create a truly sustainable product from the outset.

Beyond the physical product, consider the user experience. Intuitive designs and clear instructions promote proper use and maintenance, contributing to the product's longevity. A product that's easy to understand and use is less likely to be discarded prematurely due to frustration or misuse. A well-designed product seamlessly integrates into the user's life, becoming a reliable and valued asset rather than a fleeting purchase. This user-centric design approach should guide every stage of the product development.

Ultimately, designing for durability is a holistic approach. It requires collaboration across disciplines, from designers and engineers to manufacturers and consumers. It demands a fundamental shift in how we view products, moving away from a linear "take-make-dispose" model to a circular model of continuous use and reuse. By prioritizing durability, we can create a more sustainable future, one where our consumption patterns align with the health of the planet. The transition to this model requires not only the development of new technologies and practices but also a cultural shift towards valuing longevity over disposability. It's a challenge, but one worth embracing for a truly sustainable future.

4.2 Product Lifecycle Management

Effective Product Lifecycle Management (PLM) is crucial for achieving zero-waste goals. It's not just about recycling; it's about designing products from the outset with their entire lifespan in mind, minimizing waste at every stage, from raw material extraction to final disposal or, ideally, reuse. This requires a radical shift from the traditional linear "take-make-dispose" model to a circular approach where materials are kept in use for as long as possible.

Consider the impact of a simple plastic bottle. Its creation consumes resources, its transportation utilizes energy, and its eventual disposal often involves landfill or incineration. However, through thoughtful PLM, we can significantly reduce its environmental impact. This might involve using recycled materials in its production, designing it for easy disassembly and recycling, or even creating a closed-loop system where the plastic is repeatedly reused. This requires collaboration between designers, manufacturers, and consumers.

Implementing comprehensive PLM involves several key strategies. Firstly, design for durability is paramount. Products should be built to last, resisting wear and tear, and easily repaired or upgraded. This contrasts with the current trend of planned obsolescence, where products are deliberately designed to fail after a short period, driving up consumption and waste. Investing in robust, long-lasting products ultimately saves money in the long run, and reduces our environmental footprint. Secondly, designing for disassembly is equally crucial. Products should be easily broken down into their constituent parts, allowing for efficient material recovery and recycling at the end of their useful life. Avoid using composite materials which make separation difficult and uneconomical.

Materials selection plays a vital role. Prioritizing recycled and renewable materials reduces the demand for virgin resources and minimizes environmental damage. Life cycle assessments (LCAs) can help identify materials with the lowest environmental impact throughout their entire life cycle. These assessments consider every stage, from extraction and processing to manufacturing, use, and disposal. The results provide vital information guiding better material choices. Transparency is also key; consumers need to be informed about the materials used and the environmental impact of products to encourage responsible purchasing choices.

Extending product lifecycles requires innovative approaches. This includes designing for reuse, repair, and refurbishment. Creating a robust repair infrastructure, with readily available spare parts and skilled technicians, extends the life of products significantly. Refurbishment programs can give products a second life, reducing waste and extending their value. Product-as-a-service models, where companies offer the use of products rather than selling them outright, foster long-term responsibility for their products' environmental impact, reducing the demand for resources.

Furthermore, incorporating end-of-life management into PLM is indispensable. This involves planning for the responsible disposal or recycling of products at the end of their useful life. Working with recycling facilities and ensuring that components are easily recyclable reduces the amount of waste sent to landfill. Close collaboration with waste management professionals can lead to innovative solutions which ensure valuable materials are recovered, preventing them from ending up in landfills or polluting the environment. By integrating end-of-life planning from the design phase, the environmental impact is minimized throughout the entire lifecycle of the product. Ultimately, the goal of a circular economy cannot be achieved without taking into account the entire lifecycle of each and every product.

Finally, effective PLM requires collaboration across the supply chain. Manufacturers, designers, consumers, and waste management companies must work together to achieve zero-waste goals. This involves transparent communication, shared responsibility, and a commitment to continuous improvement. Incentivizing sustainable practices and penalizing irresponsible ones further reinforces these principles. Investing in research and development of new materials and technologies that support circularity is also essential for progress. Only through a collaborative effort can a truly circular economy that eliminates waste emerge. This holistic, collaborative approach is critical for achieving a truly sustainable future. It is only by working together that we can create a system that conserves resources, reduces waste, and protects our planet.

5 Community Action and Policy

Local governments play a crucial role. They can implement effective waste diversion programs, such as curbside composting collection, and incentivize businesses to adopt sustainable practices through tax breaks or grants for waste reduction initiatives. Furthermore, robust public education campaigns, easily accessible to all demographics, are paramount. These campaigns should focus not just on recycling, but on the entire hierarchy of waste management: refuse, reduce, reuse, recycle, and rot (composting). By making information readily available and easily understandable, communities can be empowered to make informed choices, contributing to significant waste reduction.

Effective advocacy requires a multi-pronged approach. Start by building strong relationships with local officials. Attend council meetings, participate in public forums, and engage in direct dialogue with elected representatives. Lobbying for policies that support zero-waste initiatives is critical; this includes promoting extended producer responsibility schemes, where manufacturers are held accountable for the end-of-life management of their products. Simultaneously, building a powerful grassroots movement, one that unites diverse community groups, is equally vital. Collective action amplifies voices and increases the likelihood of policy change. Consider forming coalitions with environmental organizations, community gardens, and other groups sharing similar goals.

Community composting programs offer a tangible way to tackle organic waste. These programs can take various forms, from small-scale neighborhood composting initiatives to larger-scale municipal programs utilizing industrial composting facilities. Regardless of the scale, successful programs need clear guidelines, readily accessible composting bins, and ongoing education to ensure proper composting techniques. Consider partnering with local businesses and schools to establish composting sites; this not only reduces waste but also provides valuable learning opportunities, fostering a sense of community ownership. The success of these programs rests on effective communication and continuous engagement with community members.

Waste reduction strategies must focus on the root causes of waste generation. This means addressing issues like unsustainable packaging, excessive single-use plastics, and inefficient product design. Promoting the use of reusable containers, bags, and other items reduces the overall volume of waste going to landfills. Support local businesses that prioritize sustainable packaging and bulk purchasing options. Advocate for policies that incentivize the use of reusable materials and discourage the use of single-use plastics. Remember that reducing waste at the source is far more effective than relying solely on recycling and composting. We need to change our consumption habits from a linear "take-make-dispose" model to a circular "reduce-reuse-recycle" approach.

Educational campaigns are the bedrock of any successful zero-waste initiative. These campaigns should be tailored to different age groups and demographics, employing diverse communication channels to reach a wide audience. Schools can integrate zero-waste principles into their curriculum, educating young people about the environmental impacts of waste and empowering them to become active participants in creating a sustainable future. Community workshops, demonstrations, and interactive events can teach practical skills, such as home composting and upcycling techniques, fostering a sense of community involvement. The key is to make learning engaging and accessible to all. Effective educational campaigns should equip community members with the knowledge and skills to make sustainable choices in their daily lives. It's not about telling people what to do, but about empowering them to make informed decisions. Ultimately, it is collective action, fueled by education and driven by community engagement, that will achieve the goal of zero-waste communities.

5.1 Local Initiatives

Local initiatives are the bedrock of any successful zero-waste movement. They provide tangible examples of what's possible, inspiring broader adoption of sustainable practices and demonstrating the feasibility of community-led change on a smaller, more manageable scale. These initiatives aren't just about recycling bins; they represent a fundamental shift in how communities view waste, transforming it from a problem to be solved into a resource to be managed. Success hinges on strong community buy-in and creative solutions tailored to the specific needs and resources available locally.

Consider the example of a small town in Vermont that established a community-wide composting program. Instead of relying solely on municipal services, residents actively participated by collecting food scraps and yard waste in designated bins. These were then transported to a centrally located composting facility managed by volunteers. The resulting compost was then used to enrich the town's public gardens and parks, closing the loop and creating a tangible benefit for everyone involved. This initiative not only reduced landfill waste dramatically, but also fostered a stronger sense of community ownership and responsibility. It's a brilliant example of turning a waste stream into a valuable community resource.

Community gardens often serve as excellent hubs for local zero-waste initiatives. They provide spaces for residents to grow their own food, reducing reliance on commercially produced, often heavily packaged produce. The gardens also become natural composting sites, with food scraps and yard waste being utilized to enrich the soil. Such collaborative efforts not only lessen the overall waste burden, but also strengthen local food systems, promote healthy eating habits, and bring neighbors together in a meaningful way. The shared experience fosters a sense of collective responsibility for environmental stewardship, something easily lost in larger, more impersonal settings.

Another powerful strategy involves partnering with local businesses to reduce packaging and promote reusable options. This might involve encouraging restaurants to use compostable takeout containers, working with grocery stores to offer bulk food options minimizing single-use plastics, or collaborating with local retailers to establish a system for exchanging or repairing damaged goods rather than simply discarding them. The engagement of local businesses is vital; they often possess valuable resources and insights that individuals may lack, creating synergistic partnerships benefiting the entire community. This strategy is not merely about persuading businesses to adopt sustainable practices, it's about fostering a collaborative relationship that results in mutually beneficial outcomes.

Beyond these examples, several other critical actions exist. Organizing regular community cleanups, focusing on littered areas and particularly troublesome spots like waterways and parks, is a direct, visible method of improving the local environment and fostering a collective sense of ownership of public spaces. These events not only remove litter but also raise awareness about the problem of waste, encouraging residents to adopt more responsible behaviors in their daily lives. Participation in such events creates a powerful sense of shared purpose, transforming community engagement into an active process that directly benefits everyone.

Creating educational campaigns targeted at different community groups is also essential. These should range from workshops on home composting and waste reduction to school programs teaching children about the importance of environmental responsibility and sustainable practices. This targeted approach can be tailored to resonate with specific audiences, maximizing the impact of these messages and promoting behavior change across different demographics. Children are particularly susceptible to these messages, carrying this awareness and new habits into their adult lives, creating a ripple effect that lasts for generations. For sustainable change, cultivating this understanding is a cornerstone. These educational campaigns often serve as springboards for other more involved initiatives.

Finally, effective local initiatives require strong partnerships between community groups, local governments, and businesses. This collaboration establishes a cohesive framework for implementing and sustaining zero-waste programs over the long term. It's not just about shared responsibility; it's about pooling resources, expertise, and energy to achieve common goals that benefit the community as a whole. Establishing clear communication channels is essential. This collaborative approach enables the community to address waste management issues comprehensively, proactively adapting to new challenges and maximizing the success of any initiative.

5.2 Advocacy and Policy Change

Effective advocacy for policy change requires a multi-pronged approach. It's not simply about writing letters or attending meetings; it's about building a strong, informed, and persistent movement. Start by identifying your local representatives – council members, state senators, and even federal representatives – and familiarize yourself with their stances on environmental issues. This groundwork allows you to tailor your communications to resonate with their existing priorities, increasing the likelihood of a receptive ear. Remember, persuasive arguments are grounded in facts, not emotions, so build your case with data on local waste generation, recycling rates, and the economic benefits of a circular economy.

Building coalitions is crucial. Partnering with local businesses, environmental organizations, community groups, and even schools amplifies your message and provides diverse perspectives. A united front presents a more compelling case for change than individual voices alone. Consider organizing events like community cleanups or workshops to raise awareness and engage potential allies. This hands-on approach builds community support and demonstrates the practical impact of your advocacy. Remember to document your progress; photographs of community events, data illustrating the effectiveness of local initiatives, and anecdotal evidence of individual impact all serve as potent evidence of your campaign's success.

Direct engagement with policymakers is key. Arrange meetings, write detailed proposals outlining your suggested policies, and follow up consistently. Don't underestimate the power of personal storytelling; sharing real-life examples of the challenges faced by your community due to inadequate waste management systems can be profoundly impactful. Remember, persistent communication is crucial; politicians are busy, and your message may require repeated delivery through various channels. Use social media to broaden your reach and engage a wider audience. Highlight successes and demonstrate the growing support for your cause; this momentum can exert considerable pressure on those in power.

Finally, focus on framing your advocacy within a broader narrative of economic benefit. Highlight the job creation potential of a thriving circular economy, emphasizing the opportunities in waste management, recycling, and the creation of durable, repairable products. Emphasize the long-term cost savings associated with reducing waste and investing in sustainable infrastructure, and contrast this with the escalating costs of landfill management and pollution cleanup. By showcasing the economic advantages, you can appeal to policymakers' practical concerns and demonstrate that embracing sustainable solutions isn't merely an environmental imperative but also a sound financial decision, one that strengthens the local economy and builds a more resilient community. Remember, policy change is a marathon, not a sprint. Persistence, collaboration, and a well-crafted strategy are the keys to success. Never lose sight of the ultimate goal: a cleaner, healthier, and more sustainable environment for everyone.

Crafting effective policy proposals demands a structured and meticulous approach. Begin by clearly defining the problem – for example, inadequate recycling infrastructure or lack of compost collection services. Then, propose concrete solutions backed by evidence. This might involve advocating for increased funding for waste management programs, implementing pay-as-you-throw schemes to incentivize waste reduction, or enacting regulations to promote the use of reusable containers. Ensure your proposal includes measurable objectives and timelines, enabling policymakers to assess the progress and efficacy of your suggested policies. Also, consider the potential challenges to implementation and propose solutions to overcome them. For instance, you might suggest pilot programs to test the viability of new initiatives before broader implementation, or offer solutions for handling community resistance to certain changes.

Remember, transparency is paramount. Clearly outline the projected costs and benefits of your proposals, ensuring financial projections are realistic and verifiable. Transparency inspires trust and credibility, making your proposals more palatable to policymakers and the wider community. Furthermore, make your proposals accessible and easy to understand. Avoid technical jargon and use clear, concise language to make the key points immediately evident. Consider creating visually appealing summaries alongside the detailed proposal to make it easier for policymakers to grasp your key recommendations. Remember that even small victories contribute to broader changes. If your initial proposal isn't fully adopted, advocate for the adoption of some of its aspects; this demonstrates incremental progress and sustains momentum for future advocacy efforts.

5.3 Community Composting Programs

Community composting programs offer a powerful solution to organic waste management, a significant contributor to landfills and greenhouse gas emissions. These programs actively engage the community, fostering a sense of shared responsibility and environmental stewardship, unlike many top-down approaches to waste management. Successful programs often involve a multi-pronged approach, combining convenient drop-off locations with educational outreach to promote participation and proper composting techniques. For example, some cities provide designated bins in parks or community centers, making composting accessible to residents even without dedicated home composting setups. Others offer workshops on composting methods and best practices.

The benefits extend beyond waste reduction. Community composting programs can create valuable resources like nutrient-rich compost, which is often made available to community gardeners or distributed to local farms, strengthening local food systems and reducing the need for synthetic fertilizers. This closed-loop system showcases the circular economy principles in action, transforming waste into a valuable resource for the community. Further, the shared work of collection and management can build social bonds and create a more resilient community. Some programs incorporate volunteers, allowing members to actively contribute to a shared cause.

Financial sustainability is a crucial element for long-term success.

While some programs rely on municipal funding, others explore innovative funding models. These may involve charging a small fee for drop-off service, securing grants from environmental organizations, or forming partnerships with local businesses interested in sourcing high-quality compost. A comprehensive financial plan, including realistic budgeting and potential revenue streams, is essential for securing ongoing support. Regular monitoring and evaluation are also crucial to assess program effectiveness and adapt strategies based on community needs and participation levels. Effective communication with residents is necessary to address any concerns, answer questions, and maintain ongoing community engagement.

The creation of effective community composting programs often requires navigating various regulatory hurdles and securing necessary permits. Understanding local regulations concerning waste disposal and composting is vital, alongside adhering to health and safety standards to ensure the program operates safely and legally. Collaborating with local authorities and engaging with relevant stakeholders, including waste management companies and environmental agencies, is crucial during the planning and implementation phases. This collaborative approach can facilitate smoother program implementation and long-term sustainability. Furthermore, building strong relationships with community leaders, neighborhood associations, and local businesses can help promote wider community participation and acceptance.

Effective community engagement requires more than just setting up collection points. Ongoing educational initiatives are crucial for fostering responsible participation and dispelling misconceptions about composting. Public awareness campaigns, interactive workshops, and online resources can educate residents on acceptable materials, composting techniques, and the benefits of the program. Regular feedback mechanisms—surveys, community meetings, or online forums—allow for valuable insights and improvement opportunities. Addressing concerns and misconceptions transparently and proactively builds trust and encourages continued community support. A robust communication strategy can significantly increase participation rates and the overall success of the program. Regular feedback mechanisms, such as surveys or community forums, are needed to refine the program and keep the community engaged.

5.4 Waste Reduction Strategies

Effective waste reduction requires a multifaceted approach, moving beyond simple recycling. It necessitates a fundamental shift in mindset, prioritizing prevention over remediation. We need to actively design out waste from the very beginning of a product's lifecycle, focusing on durable, repairable goods. This isn't just about individual actions; successful strategies require systemic changes, encompassing businesses, governments, and communities working collaboratively.

Consider the impact of packaging. Excessive packaging, often unnecessary and non-recyclable, contributes significantly to landfill waste. Demand reusable containers and packaging alternatives from businesses, actively choosing products with minimal or easily recyclable packaging. This simple act, repeated by many, creates significant collective impact. Lobby for policies that incentivize businesses to reduce packaging waste and penalize excessive use. Support companies that embrace sustainable packaging practices, rewarding those who prioritize environmental responsibility.

Composting, often overlooked, offers a powerful solution for organic waste. This process transforms food scraps, yard waste, and other organic materials into nutrient-rich soil amendments. Start with home composting; it's easier than you might think. Even a small-scale compost bin in your backyard can significantly reduce your household's waste. Advocate for community composting programs, working with local governments to expand access and encourage broader participation. These programs can transform organic waste into a valuable community resource.

Beyond composting, explore anaerobic digestion. This advanced technology uses microorganisms to break down organic materials in the absence of oxygen, producing biogas that can be used for energy. While requiring a larger scale investment, anaerobic digestion offers a significant opportunity to manage waste sustainably and generate renewable energy. Support initiatives that explore and implement this technology within your community. Your support, even at a grassroots level, raises awareness of the advantages.

Education is key to effective waste reduction. Raising awareness about the impacts of waste generation is essential. Support local educational initiatives that promote zero-waste practices. Teach children and adults about the importance of reducing consumption, repairing items instead of replacing them, and choosing sustainable products. Organize workshops and community events focused on waste reduction techniques, sharing practical advice and inspiring collective action. These campaigns can ripple outward, influencing behavior changes within your local community and beyond.

Remember, reducing waste is not solely about individual actions. It requires a collaborative effort. Engage with local government, advocating for policies that support waste reduction and encourage the transition to a circular economy. Support businesses that prioritize sustainability, rewarding their commitment to environmental responsibility. Become an active member of your community, organizing initiatives and partnering with others to promote zero-waste practices. Your participation, no matter how small, contributes to the larger movement towards a more sustainable future. Every action, from choosing reusable bags to advocating for policy changes, makes a difference. By working together, we can significantly reduce waste and create a healthier environment for generations to come. Small changes, consistently practiced, create lasting transformations. The collective effort will achieve systemic change. Your commitment to a zero-waste lifestyle represents a significant contribution. This proactive engagement fosters a broader community-wide commitment towards environmental responsibility. This, in essence, is the foundation of effective waste reduction strategies.

5.5 Educational Campaigns

Effective educational campaigns are crucial for transitioning towards zero-waste communities. They must move beyond simple awareness-raising and actively engage citizens in adopting sustainable practices. This requires a multi-pronged approach, combining various communication channels and interactive learning experiences. We must remember that effective education isn't simply about imparting information; it's about fostering behavioral change.

Successful campaigns should leverage diverse media to reach a wide audience. Think beyond traditional flyers and pamphlets; consider using social media platforms, podcasts, and engaging videos to showcase the benefits of zero-waste living in relatable ways. Collaborations with local influencers and community leaders can boost campaign reach and credibility, establishing trust and encouraging participation. Think of creating short, easily shareable videos demonstrating simple composting techniques or showcasing local zero-waste businesses.

Interactive workshops and community events offer a powerful way to engage directly with citizens. Hands-on activities, such as composting demonstrations, upcycling workshops, or repair cafes, can transform abstract concepts into tangible experiences. By actively participating, individuals connect with the zero-waste movement on a personal level, strengthening their commitment to sustainable practices. These interactive sessions could also involve expert speakers addressing specific topics, such as the environmental impact of different materials and recycling methods.

Targeting specific demographics is crucial. Tailor messaging to resonate with different age groups and community segments. For instance, campaigns aimed at young people might employ social media trends and interactive games, while older adults may respond better to community-based workshops and presentations. Consider the unique needs and challenges of different groups. For example, low-income households might require education on affordable zero-waste solutions or access to resources.

Integrating zero-waste education into school curricula is essential for building a future generation of environmentally conscious citizens. This should not simply be a one-off lesson but rather an integrated approach across various subjects, from science and social studies to arts and crafts. Teaching children about the importance of resource conservation, recycling, and responsible consumption from a young age creates lasting behavioral changes. Furthermore, schools could implement their own composting and recycling programs to reinforce the learning in practice.

Effective campaigns should also address common misconceptions surrounding recycling and waste management. Clear, evidence-based information can dispel myths and clarify the limitations of current recycling systems. This transparency fosters trust and encourages community members to adopt more effective waste reduction strategies, understanding that recycling is only part of the solution. Explaining the process of downcycling and the importance of reducing consumption before recycling is paramount.

Measuring campaign effectiveness is crucial. Track participation rates in workshops, social media engagement, and changes in waste generation at the community level. This data will inform future campaigns, helping to refine strategies and maximize impact. This constant evaluation and adaptation guarantee that the campaign remains responsive and relevant to the community's changing needs. Incorporating feedback mechanisms allows for continual improvement and a better understanding of the public's understanding of zero-waste principles.

Beyond disseminating information, successful campaigns should empower individuals and communities to take action. Provide clear guidelines on implementing zero-waste practices at home and in the workplace. Share resources such as composting guides, repair tutorials, and local zero-waste businesses directories. Create a supportive environment where community members can share tips, overcome challenges, and celebrate successes together. This fostering of a supportive community can increase the adoption of sustainable habits.

Finally, remember that collaboration is key. Partner with local governments, environmental organizations, businesses, and community groups to maximize reach and impact. A unified and consistent message across different platforms strengthens the campaign's credibility and increases the likelihood of achieving its objectives. This shared commitment and a coordinated approach are essential for driving widespread adoption of zero-waste principles and building a more sustainable future for all.

6 Practical Guide to Zero Waste Living

Start by honestly assessing your current waste production. Weigh your trash, meticulously separating recyclables, compostables, and landfill waste. This data provides a baseline, a crucial starting point for tracking your progress and identifying areas needing immediate attention. Understanding where your waste originates – excessive packaging, disposable items, food scraps – will illuminate the most impactful changes you can make. Remember, even small shifts can accumulate into significant reductions over time; consistency is key. Don't aim for perfection immediately; focus on steady, sustainable progress. This self-audit is your first step towards a truly transformative lifestyle.

Next, focus on reducing food waste, a significant contributor to landfills. Plan your meals meticulously, utilizing leftovers creatively and storing food properly to extend its shelf life. Compost food scraps diligently; this not only diverts waste but enriches your soil, fostering a more sustainable gardening practice. Learn to preserve food through methods like canning, freezing, or pickling. These techniques extend the usable life of produce, minimizing trips to the grocery store and the subsequent packaging waste. Consider joining a community-supported agriculture (CSA) program; you'll receive fresh, seasonal produce while reducing the environmental impact of large-scale farming and distribution.

The "Refuse, Reduce, Reuse, Recycle" hierarchy is your guiding principle. Refuse single-use plastics, unnecessary packaging, and free promotional items whenever possible. Carry reusable bags, water bottles, and coffee cups. Reduce consumption by prioritizing experiences over material possessions, repairing items instead of replacing them, and choosing durable, long-lasting products. Reuse containers, jars, and bags for storage or repurpose them creatively. Recycle diligently, adhering to your local guidelines and knowing the limitations of your community's recycling program. Prioritize the first three "Rs"—Refuse, Reduce, and Reuse—before resorting to recycling. Remember, recycling is a last resort, not a solution to overconsumption.

Home composting is surprisingly simple and incredibly rewarding. A variety of methods exist, from simple backyard bins to sophisticated indoor composting systems. Select a method suited to your space and lifestyle. Understanding the composting process—balancing brown materials (leaves, twigs) with green materials (fruit, vegetable scraps)—is crucial. Maintain proper aeration and moisture levels to prevent unpleasant odors and attract pests. The resulting compost enriches your garden soil, reduces your reliance on chemical fertilizers, and helps you close the loop on your food waste. The rewards go beyond just reducing waste; you actively participate in creating healthy, fertile soil.

Beyond individual actions, community involvement is essential. Participate in local cleanups, support businesses with sustainable practices, and advocate for policies that prioritize waste reduction. Connect with community gardens or composting programs; these collaborative efforts magnify the impact of individual actions. Seek out zero-waste stores, participate in workshops, and educate others about sustainable practices. Remember, building a zero-waste community requires collective effort. Your actions inspire others, creating a ripple effect of positive change that extends beyond your own home. Every conscious choice contributes to a healthier planet, a better future. This ongoing commitment and participation within your community will transform not only your own habits, but inspire positive change on a larger scale. That is the true power of a zero-waste lifestyle.

6.1 Home Composting Methods

Home composting offers a fantastic way to divert organic waste from landfills, reducing methane emissions and enriching your garden. It's surprisingly simple, even for beginners, and the rewards—nutrient-rich soil for your plants—are substantial. Several methods exist, each with its own advantages and disadvantages, depending on your space, climate, and composting style.

The simplest method is the backyard compost heap. This involves creating a pile of organic materials—fruit and vegetable scraps, yard waste, coffee grounds—in a designated area of your yard. Regular turning is crucial for proper decomposition, ensuring adequate aeration and promoting a healthy microbial ecosystem. You can construct a simple three-sided enclosure from readily available materials like pallets or chicken wire to keep the pile contained and improve its overall structure. This method works best in areas with sufficient space and moderate climates.

For those with limited outdoor space, a compost bin provides a more contained and aesthetically pleasing alternative. Compost bins come in various sizes and materials, ranging from simple plastic bins to more sophisticated designs with built-in aeration systems. These bins generally offer better control over the composting process and prevent the mess and odors that can sometimes accompany open compost heaps. Choose a bin size appropriate for your household's waste generation. Smaller bins require more frequent emptying, while larger ones may require more effort in turning the contents.

Vermicomposting, or composting with worms, is another excellent option, particularly suitable for apartment dwellers or those with minimal outdoor space. This method utilizes red wiggler worms, which efficiently break down organic matter, producing a nutrient-rich worm castings, highly valued in gardening. Vermicomposting typically involves a layered system in a container, with layers of organic material and bedding materials alternating. Regular monitoring of moisture levels is crucial for maintaining a healthy worm population. This system requires less space but necessitates regular attention to ensure proper moisture and worm health.

Regardless of the method you choose, understanding the basic principles of composting is critical. Maintain a balance of "green" materials (nitrogen-rich, such as fruit and vegetable scraps) and "brown" materials (carbon-rich, such as dried leaves and twigs). A good ratio is roughly 1:2, green to brown. This ensures proper decomposition and avoids unpleasant odors. Also, keep the compost pile moist, but not soggy. Overly wet compost can become anaerobic, leading to foul smells and slow decomposition, so good drainage is essential. Regularly turning the compost will greatly accelerate the decomposition process.

Remember, not all organic materials are suitable for composting. Meat, dairy products, oily foods, and diseased plants should be avoided, as they can attract pests and create unpleasant odors. Similarly, certain weeds, like bindweed, might re-sprout in your compost and spread to your garden. If you are unsure about a specific material, it's always best to err on the side of caution and dispose of it separately. Properly managed composting reduces your waste output, enriches your soil, and contributes to a healthier environment. The resulting compost is a valuable resource for your garden, providing essential nutrients for healthy plant growth and improving soil structure. By incorporating these tips into your composting practices, you will enhance your composting success and reap significant rewards. You'll be creating a closed-loop system where waste becomes a valuable resource, demonstrating a practical commitment to a circular economy and sustainable living. Through this simple act, you actively participate in reducing your environmental footprint and embracing the transformative power of recycling. Composting is more than just waste management; it's an investment in the health of your garden and the planet. Embrace the process and enjoy the benefits. This active participation in the circular economy will inspire you to take further steps in your zero-waste journey. Beyond composting, consider expanding your sustainable practices by reducing food waste, reusing items whenever possible, and diligently recycling. These integrated efforts collectively contribute to building a more sustainable future for our communities. Ultimately, the success of composting, and indeed your broader zero-waste lifestyle, relies on consistent effort and a commitment to continuous learning and improvement. Don't be discouraged by initial challenges—the rewards are significant and far outweigh any initial hurdles you might encounter. Your commitment, however small, directly contributes to building a more sustainable future.

6.2 Reducing Food Waste

Food waste represents a significant challenge in our pursuit of zero-waste communities. Globally, a considerable portion of edible food ends up in landfills, contributing to greenhouse gas emissions and resource depletion. Tackling this issue demands a multi-pronged approach, starting with mindful shopping habits and extending to innovative storage solutions.

Effective planning is crucial. Before you even step into a grocery store, create a detailed meal plan for the week. This simple act prevents impulsive purchases that often lead to uneaten food. Utilize a shopping list meticulously crafted from your plan; sticking to it minimizes the risk of acquiring excess ingredients. Regularly check your refrigerator and pantry, taking stock of what you already have. This prevents accidental duplication and reduces the likelihood of items expiring before use.

Proper storage techniques play a vital role in extending the shelf life of perishable goods. Refrigerate items promptly upon arrival, ensuring they're stored at the correct temperature. Use airtight containers to maintain freshness and prevent spoilage. For fruits and vegetables, consider methods like storing them separately to prevent ethylene gas buildup which accelerates ripening and decay. Learning about the best storage practices for different types of produce can drastically extend their usability.

Understanding "first in, first out" (FIFO) is essential for preventing food waste. Place newer items behind older ones in your refrigerator and pantry. This ensures that older products are used before they expire. Regularly rotate your stock, bringing forward items nearing their expiry dates to increase the chances of consumption. Visual cues can help; label containers with dates to easily track how long items have been stored.

Freezing is a powerful ally in the fight against food waste. Many fruits, vegetables, and even cooked meals can be frozen to preserve them for later use. Portioning food before freezing optimizes space and makes it easier to use smaller amounts without defrosting the entire batch. Freezing doesn't just prevent spoilage; it preserves nutrients and flavors.

Composting is a fantastic way to manage food scraps that aren't suitable for consumption. Even if you have limited space, indoor composting options exist. Vermicomposting, using worms to break down organic matter, is a particularly efficient method, producing nutrient-rich compost for your garden. Learn about the specific requirements for your chosen composting method, including the necessary ratio of browns (e.g., dried leaves) to greens (e.g., food scraps).

Transforming leftovers is an art that minimizes waste and expands culinary creativity. Cooked meals can be repurposed into new dishes, reducing the need for additional ingredients and cooking time. A simple leftover chicken can become the base of a hearty soup or salad. Vegetable scraps can find new life in homemade broths, adding depth and flavour to future meals.

Don't underestimate the power of creativity. Get inventive with your use of almost-expired ingredients. Overripe bananas can be used in smoothies or baked goods. Slightly wilted vegetables can be added to

soups or stir-fries. Think outside the box; you might be surprised at the delicious results.
Sharing surplus food with friends, neighbors, or community organizations is a simple yet profound way to address food waste. Local food banks and shelters greatly appreciate edible donations. Consider arranging a regular swap with neighbours, exchanging extra produce or home-cooked meals. These gestures not only reduce waste but also foster community ties.
Finally, remember that reducing food waste is a journey, not a destination. Embrace imperfection, learn from your mistakes, and celebrate your successes. Small, consistent actions over time make a significant impact. Through mindful planning, proper storage, creative repurposing, and community sharing, we can significantly lessen our collective footprint and contribute to a more sustainable future for all.

6.3 Refuse, Reduce, Reuse, Recycle

The core of minimizing your environmental footprint lies in understanding and diligently applying the four Rs: Refuse, Reduce, Reuse, and Recycle. It's a hierarchy; refusing is always preferable to reducing, reducing to reusing, and reusing to recycling. Prioritizing refusal means actively choosing not to acquire items you don't truly need. This involves critically evaluating marketing strategies, resisting impulse buys, and asking yourself if you really need something, or if you simply want it. Consider the true cost – not only financial but environmental – before purchasing any product. This conscious decision-making process is the foundation of a zero-waste lifestyle.
Reducing consumption is the next crucial step. Once you've refused un-

necessary items, concentrate on minimizing what you do buy. Buy in bulk when possible to reduce packaging waste. Choose products with minimal or recyclable packaging, opting for items with less packaging whenever feasible. Support businesses committed to sustainable practices. Remember that every purchase has an environmental impact; by reducing your consumption, you reduce that impact significantly. This isn't about deprivation; it's about mindful, intentional purchasing.

Reusing items extends their lifespan and dramatically cuts down on waste. Repurposing old jars for storage, using cloth bags instead of plastic ones, and mending clothes instead of replacing them are all examples of reuse. Get creative; find new ways to utilize items before discarding them. The internet provides an endless source of inspiration for repurposing; exploring this realm can unlock a surprising amount of creativity and resourcefulness. Consider donating or selling unwanted items rather than throwing them away. This step keeps usable goods in circulation, reducing the demand for new products.

Recycling, the final R, is undeniably important but is only effective if done correctly and should always be considered a last resort after refusing, reducing, and reusing. Understand your local recycling guidelines thoroughly. Many materials aren't truly recyclable in your area, leading to contamination and ultimately landfill. Familiarize yourself with your municipality's recycling program to ensure you're properly sorting materials. Furthermore, remember that even recyclable materials consume energy and resources during the recycling process. Prioritizing the first three Rs minimizes the reliance on recycling, making the entire process more effective and sustainable.

Properly sorting and preparing recyclables for collection is vital to maintaining the integrity of recycling streams. Rinse containers thoroughly to prevent contamination. Flatten cardboard boxes to save space. Keep separate containers for different materials—paper, plastic, glass, metal. This seemingly small step ensures that recyclables don't get contaminated and can be successfully processed. Additionally, research local initiatives that offer alternative recycling methods or advanced recycling technologies such as chemical recycling or enzymatic depolymerization for hard-to-recycle plastics.

Beyond these four Rs, consider the broader implications of your choices. Support local farmers' markets, reducing transportation emissions and supporting sustainable agriculture. Purchase secondhand goods whenever feasible, giving items a second life and reducing the demand for new production. These actions represent a shift from a linear consumption model—take, make, dispose—to a circular economy, where resources are utilized to their fullest potential, reducing waste, pollution, and our reliance on finite resources. The transition requires conscious effort, but the collective impact is significant. Every small change contributes to creating a more sustainable future.

7 Conclusion: Building a Sustainable Future

Embracing a zero-waste lifestyle isn't merely a trend; it's a fundamental shift in how we interact with the planet. This journey, as we've explored throughout this book, requires a conscious effort to reconsider our consumption habits, understand the limitations of current recycling systems, and actively participate in building circular economies. The transformation demands a collective commitment, a global movement driven by individuals, communities, and governments alike. Our individual actions, while seemingly small, collectively create a powerful wave of change.

This collective action transcends geographical boundaries. Global collaboration is crucial. International agreements and standardized practices are needed to streamline recycling processes, facilitate the sharing of best practices in waste management, and create a global network dedicated to resource efficiency and waste reduction. Shared data and research can be leveraged to identify the most effective strategies for different environments and socio-economic contexts. Such collaborations foster innovation, enabling the development of new technologies and solutions that address challenges unique to certain regions.

Consider, for instance, the exchange of knowledge regarding anaerobic digestion techniques between developed and developing nations. Developing countries often lack the infrastructure for widespread recycling; adopting anaerobic digestion offers a feasible alternative for managing organic waste, producing biogas as a renewable energy source. Simultaneously, developed nations can learn from innovative waste management strategies implemented in resource-constrained environments, showcasing the ingenuity and adaptability of zero-waste practices. The synergistic effect of these exchanges fuels global progress towards a sustainable future.

Beyond international cooperation, the success of a zero-waste future hinges on the active participation of local communities. Community-led initiatives such as composting programs, repair cafes, and collaborative consumption models can dramatically reduce waste at a grassroots level. These initiatives foster a sense of shared responsibility, creating a culture of sustainability that extends beyond individual households. Moreover, local policies play a vital role in supporting these community efforts. Incentivizing waste reduction, promoting composting infrastructure, and regulating packaging materials can create an environment where zero-waste living becomes the norm rather than the exception.

Furthermore, effective educational campaigns are paramount. Promoting awareness regarding the environmental and economic benefits of zero-waste practices is crucial for encouraging widespread adoption. Targeted educational programs, particularly for younger generations, can foster a deep-seated understanding of the interconnectedness between our consumption habits and environmental health. This proactive approach instills a culture of responsibility and ensures the long-term success of zero-waste initiatives. We must equip future generations with the knowledge and tools to navigate a sustainable future. Ultimately, building a sustainable future requires a multifaceted approach. It's not about perfection, but about consistent progress. Every small step we take – from reducing our reliance on single-use plastics to composting our food scraps – contributes to a larger movement towards a healthier planet. The journey may seem daunting, but the rewards are immense. A future characterized by reduced waste, efficient resource utilization, and a thriving environment is not a utopian dream; it's an achievable goal. Our commitment to conscious consumption, coupled with global collaboration and local community action, lays the foundation for a truly sustainable world. It is a world where recycling isn't just a process, but a cornerstone of a vibrant, circular economy that prioritizes sustainability over short-term convenience. Remember, the path toward zero waste is a continuous process of learning, adapting, and innovating; a journey that calls for our collective action today, for a better tomorrow. Let's embrace the challenge and build a future where sustainability is not just a goal, but a way of life.

7.1 The Path Forward

The journey towards zero-waste communities isn't a sprint; it's a marathon demanding consistent effort and collective action. We've explored the challenges, understood our consumption patterns, and learned the limitations of traditional waste management systems. Now, we must build upon this foundation, transforming knowledge into tangible, sustainable change. Individual actions, amplified by community initiatives and supportive policies, are the driving force behind this transformation. It's about embracing a mindset shift—from disposability to durability, from linear consumption to circularity.

This isn't about perfection; it's about progress. Every small step—from refusing single-use plastics to meticulously composting food scraps—contributes to a larger impact. Remember the principles of the circular economy: design products for durability, prioritize repair over replacement, and find innovative ways to reuse materials. Think creatively; solutions exist for almost every waste stream.

Communities play a vital role. Organize local workshops to teach composting techniques or advocate for better municipal recycling programs. Lobby your local representatives to support policies that incentivize waste reduction and invest in infrastructure for efficient waste processing. Pressure corporations to adopt more sustainable packaging and production methods. Don't be silent; your voice matters in shaping a more sustainable future.

Collaboration is key. Sharing knowledge, resources, and experiences across communities fosters a sense of collective responsibility. Networking with other zero-waste advocates, both locally and globally, expands your influence and accelerates progress. By working together, we can create a movement that pushes for systemic change and promotes the adoption of zero-waste principles on a larger scale.

Furthermore, ongoing education is paramount. Continue to learn about new technologies and innovations in waste management. Stay updated on the latest research and best practices. Engage in discussions and share your experiences to encourage others to join the movement. We must remain adaptable and innovative, refining our strategies as we learn and grow.

Ultimately, the path forward requires a fundamental change in our relationship with resources. It necessitates a conscious shift away from a culture of disposability towards one of mindful consumption, repair, and reuse. This isn't merely about reducing waste; it's about building a more resilient, equitable, and sustainable future for generations to come. Our collective commitment to this vision will determine the success of our efforts.

The path forward isn't a clearly defined road map; it's a journey of continuous learning and adaptation. Embrace the challenges, celebrate small victories, and never lose sight of the larger goal: a world where waste is minimized, resources are valued, and communities thrive within a sustainable ecosystem. Remember, the power to create this future lies within each of us. We have the knowledge; now, let's translate it into action. Let's transform our individual choices into collective impact, thereby ensuring a healthier planet for all.

The transition to a zero-waste lifestyle is not just an environmental imperative; it's an opportunity to create stronger, more connected communities. By actively engaging in local initiatives, sharing knowledge, and advocating for policy changes, we forge a path towards a shared vision of sustainability. This isn't just about reducing individual waste; it is about building a more resilient and equitable society. This collective action strengthens social bonds, builds community pride, and fosters a sense of shared responsibility in creating a sustainable future for all.

This collective endeavor is not without its hurdles. Resistance to change, limited resources, and complex regulatory frameworks can all pose significant challenges. Yet, by remaining persistent in our advocacy and collaborative efforts, by educating and inspiring others,

and by celebrating our successes along the way, we can overcome these obstacles. Progress, however incremental, is progress nonetheless.
The true measure of success lies not just in the reduction of waste itself, but in the transformation of our mindsets and systems. By fostering a culture of sustainability, innovation, and collaboration, we create a future where zero-waste is not just a goal, but a reality. A future where our communities are vibrant, resilient, and in harmony with the environment. This future demands persistent effort, collective action, and an unwavering commitment to sustainable practices. Let's embrace the challenge and build the future we envision. The path forward is clear; it's up to us to walk it.

7.2 Global Collaboration

Global collaboration is paramount for achieving a truly zero-waste future. No single nation or community can solve this complex challenge alone; a coordinated international effort is essential. The sheer scale of waste production, coupled with the interconnectedness of global supply chains and environmental systems, demands a unified approach. Without global cooperation, even the most ambitious local initiatives will remain limited in their impact. Successful strategies will require an exchange of best practices, shared resources, and a concerted commitment to harmonizing policies across borders.
Effective collaboration necessitates open communication and data sharing between nations. Transparency in waste management practices, including the precise amounts of waste generated, recycling rates, and the composition of waste streams, is crucial for informed decision-making. This shared data can inform the development of ef-

fective strategies and allow for accurate monitoring of progress towards global zero-waste goals. Furthermore, a standardized approach to waste classification and measurement is vital for comparing data and setting meaningful targets across different regions.

International agreements and collaborative research programs are essential drivers of this global effort. Collaborative research can help identify innovative technologies and best practices for waste reduction and resource recovery. These programs should encompass a wide range of disciplines, including material science, engineering, and social sciences, to address the multifaceted aspects of waste management. The sharing of research findings and technological innovations across countries will accelerate the development and deployment of effective solutions.

International funding mechanisms are equally important. Developed nations should support developing countries in their efforts to implement zero-waste strategies, providing financial and technical assistance. This includes investing in the development of waste management infrastructure, training personnel, and promoting the adoption of sustainable practices. Access to financing is particularly critical for developing countries that may lack the resources to implement large-scale waste management programs.

Beyond financial support, knowledge transfer plays a crucial role. Developed nations should actively share their knowledge and experience with developing countries, providing technical expertise and training to support local capacity building. This includes sharing best practices in waste reduction, recycling, composting, and other aspects of waste management. Regular exchange programs, workshops, and online resources can facilitate this knowledge exchange, ensuring the successful implementation of sustainable waste management practices worldwide.

A key aspect of global collaboration involves fostering a sense of shared responsibility. The problem of waste generation isn't confined to any single country; it's a global challenge requiring a global solution. Therefore, promoting international awareness and education is essential. This includes raising public awareness about the environmental and economic impacts of waste, encouraging responsible consumption and production patterns, and promoting the benefits of adopting zero-waste lifestyles globally.

Furthermore, incentivizing sustainable practices through international trade agreements and policies is crucial. For instance, incorporating sustainability criteria into international trade agreements can encourage the production and consumption of environmentally friendly goods. These policies could incentivize the use of recycled materials, support businesses adopting circular economy models, and penalize the production and import of goods with high environmental footprints.

Ultimately, achieving a zero-waste world requires a profound shift in global attitudes and practices. This involves moving away from a linear "take-make-dispose" model towards a circular economy that prioritizes resource efficiency, waste reduction, and the responsible management of resources. This transition necessitates not only technological advancements but also systemic changes in consumption patterns, production processes, and policy frameworks.

The journey towards a zero-waste future is a collective endeavor, demanding collaborative effort, knowledge sharing, and a global commitment to sustainability. Only through concerted international cooperation can we hope to effectively address this critical global challenge and create a truly sustainable planet for future generations. This requires not only individual action, but a concerted effort by governments, businesses, and communities worldwide, working together to reshape the way we consume and manage our resources. The success of this endeavor hinges on our collective willingness to embrace change and work collaboratively towards a shared vision of a sustainable future, a future free from the burden of excessive waste. This shared responsibility is the cornerstone of genuine, lasting, and impactful change.

www.ingramcontent.com/pod-product-compliance
Lightning Source LLC
LaVergne TN
LVHW022022200926
848481LV00007B/2487
* 9 7 9 8 3 4 9 3 7 8 5 5 3 *